Quantum Physics Mad

Quantum Phys...

Easy

The Introduction Guide for Those Who Flunked Math and Science in Plain, Simple English

Donald B. Grey

Quantum Physics Made Easy

Bluesource And Friends

Introduction

Chapter 1: What Is Quantum Physics, and Why Should I Learn It?

Chapter 2: Quantization and the Uncertainty Principle

Chapter 3: Waves and Particles and the Double Slit Experiment

Chapter 4: Quantum Non-Locality and the Bohr-Einstein Debates

Chapter 5: Quantum Entanglement and Teleportation

Chapter 6: Quantum Superpositions and Schrödinger's Cat

Chapter 7: String Theory and the Theory of Everything

Chapter 8: Black Holes and the Mystery of Quantum Gravity

Chapter 9: Other Interesting Facts

    Pilot Wave Theory

    Quasars

    Mini-Facts

Conclusion

Bluesource And Friends

This book is brought to you by Bluesource And Friends, a happy book publishing company.

Our motto is **"Happiness Within Pages"**
We promise to deliver amazing value to readers with our books.

We also appreciate honest book reviews from our readers.

Connect with us on our Facebook page
www.facebook.com/bluesourceandfriends and stay tuned to our latest book promotions and free giveaways.

Don't forget to claim your FREE books!

Brain Teasers:

https://tinyurl.com/karenbrainteasers

Harry Potter Trivia:

https://tinyurl.com/wizardworldtrivia

Sherlock Puzzle Book (Volume 2)

https://tinyurl.com/Sherlockpuzzlebook2

Also check out our best seller book

https://tinyurl.com/lateralthinkingpuzzles

Introduction

Congratulations on getting *Quantum Physics Made Easy*, and thank you for doing so. Quantum physics is a realm that seems unreachable and beyond understanding to many people, but it's actually an extremely fascinating branch of science. A lot of quantum physics is still largely undiscovered or unexplained and that's what makes it so fascinating! In this book, we've done our best to include and explain the most interesting and most common of concepts within the veil of quantum physics, with the goal of arming the reader with useful (and enviable) knowledge; whether you're reading for self-betterment, understanding, or bragging rights. If we do our job correctly, the reader will emerge with a newfound understanding of the workings of the universe and everything around us.

One of the most compelling draws of the sciences for many people is the potential of discovering something that was not known before. Whether someone's doing it for fame, for fortune, or just for the fun of it, discovering something new, leaving your own personal mark for the rest of humanity's time in the universe, is a tempting prospect for many. How would you feel about naming a star, and for others to know that you named it? That star would be visible in the sky for the rest of your lifetime, and more than likely for your great-great-great-grandchildren's lifetimes. Your discovery would be immortalized above for the life of the star.

The following chapters will discuss everything you would ever want to know about quantum physics, from its founding ideas, concepts, famous figures, and mysteries, to myths and fun facts, all in language that's easy for the beginner to understand. The first chapter deals with the what and why of quantum physics, while the rest go over specific concepts in more detail.

There are plenty of books on this subject on the market, so thank you again for choosing this one! Every effort was made to ensure it is full of as much useful information as possible - please enjoy!

Chapter 1: What Is Quantum Physics, and Why Should I Learn It?

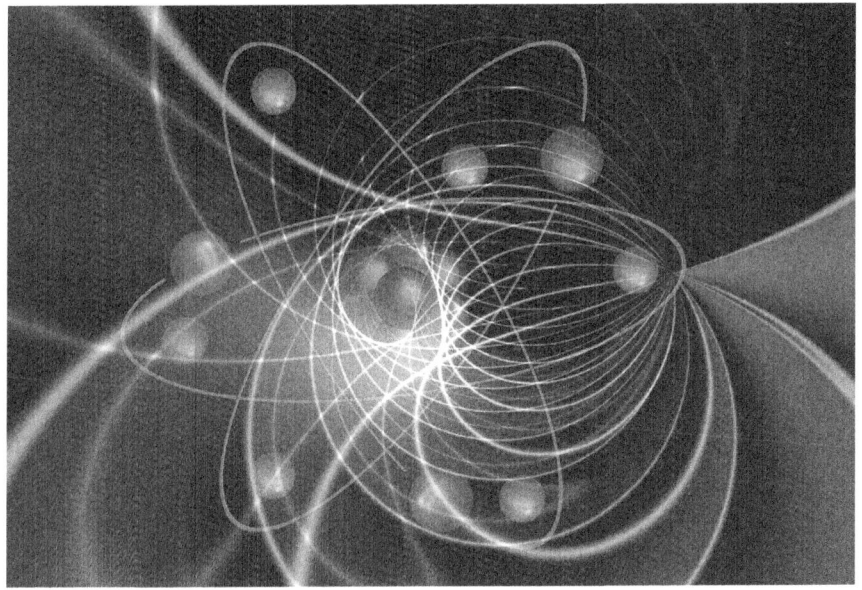

Since you've gone and paid your own hard-earned money for this book, you obviously have your own reasons for wanting to purchase and read it. However, you may not realize how intertwined the theories and nature of quantum physics can be in our everyday lives. Learning about quantum physics can help improve your understanding of sci-fi television shows and movies, deepen your understanding of space, the final frontier, and, at its simplest level, enrich your mind and daily life. Quantum physics and physics in general describe the way the world around us works on a microscopic level. Although it's full of technical jargon and mathematical equations, it's not as hard to understand as you might think, and this book aims to teach you all that you need to know. (At the very least, looking incredibly smart and intellectually advanced at your high school reunion should be on everyone's to-do list.)

When the average person hears the term *quantum physics*, however, their eyes immediately glaze over. It's a topic widely regarded to be above the reach of the average joe; a glamorous, unreachable concept lassoed only by the intellectually superior or by experts of the sciences. However, even quantum physics can be explained in an understandable way. It's also important to remember that there are some aspects of quantum mechanics (another term for quantum physics) that are still being fleshed out and discovered and that the universe still holds many mysteries.

The word "quantum" translates from Latin to mean "how much." Quantum Physics, by definition, is a theory describing how matter acts at its smallest observable scale: atoms and subatomic particles. If you're reading this book, you probably already know that all observable matter in the universe is made up of atoms. We won't be touching much on subatomic particles (the particles that are smaller than atoms) here – doing so would necessitate writing another book entirely! – but we will frequently be referring to atoms, electrons, and photons. To rehash what you may or may not remember from school: atoms are made up of protons (positively charged particles, or +), neutrons (particles without a charge, or =), and electrons (negatively charged particles, or -). The number of protons, neutrons, and electrons in the atom decides what element the atom will manifest as, and how it will behave. For example, the element Hydrogen is the simplest element that we've observed in the known universe, having only one proton and one electron. Hydrogen is also the most common element in observable existence, and studies of Hydrogen itself have helped fuel many of quantum physics' discoveries.

The difference between quantum physics and normal physics has to do with scale. Standard (or classical) physics deals with how matter acts in the observable universe; this is referred to as a macroscopic (in layman's terms, large) scale. Quantum physics, in comparison, works at a microscopic (small) scale. (If you have trouble remembering this distinction, just subtract the -scale from both words. If you do that, you get "macro," which means large, and "micro," which means small.)

Quantum physics came to be largely because of things that classical physics couldn't explain, and largely through mathematical concepts. Quantum physics, at its core, tends to be more theoretical than not: this means that there are a lot of aspects of quantum physics that we can't readily observe or provide proof of in the universe, but instead can show through mathematical equations to be true (or very likely). You'll be seeing some mathematical functions in this book, because quantum physics largely relies on math, but don't let that scare you away – we'll be explaining them in ways that any reader can understand.

One of the strangest parts of quantum physics is that, even though classical physics came first, quantum physics can be used to explain most of classical physics' phenomena on a large scale. The reverse is not true, though – classical physics cannot necessarily be used to explain the phenomena of quantum physics. At the most basic level, the main differences between classical and quantum physics boil down to four things: wave-particle duality, the uncertainty principle, quantum entanglement, and quantization. We will explain all of these concepts within the book.

Part of the theoretical nature of quantum physics is that it has a lot to do with probability. Whereas in normal physics, we can say, "that tree is right next to the fence over there," quantum physics tends to instead say, "there's a 50% chance that that tree is next to the fence, and a 50% chance that it's actually next to the fountain." Since we often can't observe things in quantum physics, we rely on this probability along with equations to explain what's going on at levels that we can't see (or can't measure).

You may have heard of Einstein's theory of relativity, which he proposed in 1905. Surely, you've heard of Albert Einstein, one of the greatest minds in history, and his formula, E=mc². However, the theory of relativity was actually his greatest brainchild. The theory of relativity is twofold: it contains both the theory of general relativity and the theory of special relativity. General relativity states that time, space, and light can be curved or distorted by acceleration and gravity. Special relativity states that light *always* travels at a fixed speed – 186,000 miles per second – no matter how fast the viewer is moving. To use a well-known example: if a person was holding a flashlight in front of them, but they were moving at the speed of light themselves, they would see a beam of light shining out from the flashlight as normal. However, to someone who was stationary, the flashlight would never appear to turn on, as the light would appear to be traveling at the same speed as the person with the flashlight.

Seems strange, right? Well, even though Einstein's theory was met with a lot of criticism when he first proposed it, today's scientists have proven many aspects of the theory, and it's widely accepted. One part of this theory that scientists have proven and observed, and that they use in research today, is that light tends to distort through gravity as it passes by a large object in space – this is called gravitational redshift, and it slightly alters the color of light to be redder. Also, congratulations – you just learned your first theory of physics!

Chapter 2: Quantization and the Uncertainty Principle

Black-body radiation refers to the way an object releases electromagnetic energy due to heat. A "black body," essentially, is an object that absorbs and releases all frequencies (colors) of light. This "black body" will change color when heated, releasing different frequencies of light by temperature. Within classical physics, there was a law called the Rayleigh-Jeans Law that could demonstrate how this energy would change with heat – however, it was only accurate at low frequencies, and as the frequency got higher (or closer to violet on the light spectrum), it would gradually spiral into infinity, becoming more and more inaccurate. This was called the Ultraviolet Catastrophe. There was also a law called Wien's law which worked the opposite way – while it described the emission of thermal light correctly at high frequencies, it became inaccurate at low frequencies (the red end of the spectrum). This was called the Infrared Catastrophe. Thus, there was a need to find rules for quantum physics that would explain this phenomenon.

Quantum Physics Made Easy

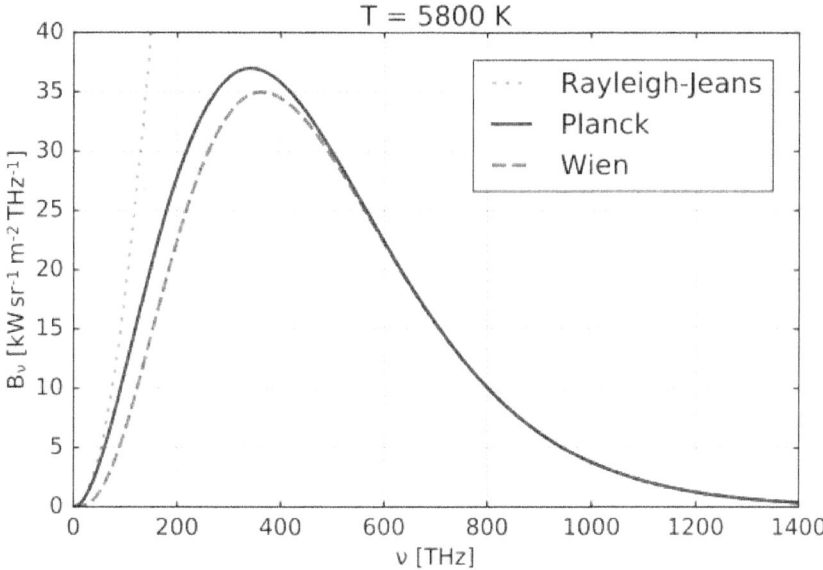

At that point in time, no one could accurately explain how this worked – that is, until Matthew Planck, in 1900, looked at it in an entirely new way. He proposed that, instead of working on a gradual spectrum of emission, electromagnetic radiation was released in "chunks." This is where quantization comes into play – this experiment was one of the earliest examples of quantization of energy. Planck theorized that, at certain points, different "chunks" or "quantities" of energy would be released – imagine, instead of a line graph with a smooth curve, a line with steps or bumps instead. These are the "chunks." Planck came up with a number that described the way this energy played out in chunks. That number is referred to as Planck's Constant, and it equals $6.62607004 \times 10^{-34}$ J/s (J/s stands for Joules per second, and Joules is a unit for measuring energy). I know that's a scary number, but the equation we're using it for is relatively easy. To find the energy of the photon (the unit of light) being released by the black body, one just needs to multiply the frequency (the color of the light) by Planck's constant. This results in this equation:

$E = hf$

In this equation, E stands for the energy of the photon (what we're solving for), h is Planck's constant, and f is the frequency of the photon. Planck's constant actually defines this "chunk" between frequencies and has to do with the threshold that the frequency of the light needs to be in order to release more energy (or more photons) from the black body. Planck didn't even know at the time what a groundbreaking discovery this was – he himself considered the quantization of this energy to be only a "formal assumption" on his part – something he threw in just to make his theory make sense mathematically. However, he was later proven correct on his assumption by multiple great minds– Albert Einstein among them – and even won a Nobel Prize for the discovery in time.

Here's another analogy to put quantization into perspective – one that we'll reuse later in the book, too. Imagine a wavy line or a guitar string. Now, turn that line into a circle. It's impossible to add less than one whole bump to that circle – if you only added half of a bump, the ends of the line would no longer meet in a circle. In order to increase the energy of this circle, you have to add one whole bump at a time to it. This is actually how an electron is believed to act – it's thought to exist as a wave circling an atom's nucleus. Planck's constant defines this quantization of the wave because, in order to increase the energy of the electron, you need to add an entire bump or a multiple of Planck's constant.

He didn't know it at the time, but Planck's constant would find far more widespread solutions than just the black body problem. In 1927, the physicist Werner Heisenberg first introduced a concept called the uncertainty principle, which was the next core concept to define quantum physics. His theory proposed that, on a quantum scale (meaning very small, such as atomic size), one cannot find the position and velocity of an object at the same time – ever. If one gets more specific with one side – say, they try to measure the velocity of an object more precisely – the position of the same object becomes even more uncertain. This is a theory that has no basis whatsoever in classical physics, since it's very hard to measure uncertainty at a macroscopic scale (for example, your eyes will tell you exactly where a car is at any given time, and a radar gun can tell you its speed. A series of calculations can tell you both).

Planck's constant is used as a foundation within the uncertainty principle. The full formula has to do with standard deviation and statistics, so we don't dive too deeply into it here, but it looks like this:

$$\sigma_x \sigma_p \geq \frac{\hbar}{2}$$

Where $\hbar = \dfrac{h}{2\pi}$

In the above equation, h is Planck's constant, σ_x is the standard deviation of the object's position, and σ_p is the standard deviation of the object's momentum. This means that the change in one part of the equation will increase as the other part is decreased – giving us the uncertainty principle.

To make this theory makes sense, one must remember that, in quantum physics (and everywhere), particles can exhibit features of waves and vice versa. Imagine the object you're looking for as ripples on a lake (ripples on water make a great analogy for waves because they *are* waves). It's pretty easy to see where the ripples are going, and we can measure the velocity of the wave by checking the distance between ripples. However, a ripple is long, and it gets wider as it moves– it's hard to describe exactly where the ripples are at a given moment in time with just a number. However, if we have a particle instead of a wave – imagine a pebble this time – it's pretty easy to describe where the pebble is in space. However, we can't tell which direction the pebble is moving, or what its velocity is. Make sense? Well, if it does, here's another crazy concept to throw on your plate – speed and position have no relevance in quantum physics. That's right – neither position nor momentum actually means anything in the quantum world. We'll cover this a bit more in chapter six, but it's because quantum particles exist in many places and states at the same time. This is the gist of the uncertainty principle. Also, keep in mind that when we observe a particle at a quantum level, we irrevocably alter the way it behaves, making it impossible to view quantum particles in their natural state. We'll go over this later, too.

Chapter 3: Waves and Particles and the Double Slit Experiment

The duality of waves and particles is one of the defining principles of quantum mechanics (we've touched on it a little already), and it's something you'll see again and again in the future and as you read this book. According to classical physics, something is either a "particle" (like an atom) or a "wave" (such as sound waves), but not both. The creators of quantum physics quickly realized that that wasn't going to work, as some things (actually, nowadays, scientists acknowledge that pretty much all of the matter in the universe exhibits wave-particle properties, including protons, neutrons, electrons, atoms, photons, and even some molecules) exhibited the properties of both particles and waves. The wave-particle duality was also one of the next definitions of quantum mechanics to be discovered and accepted.

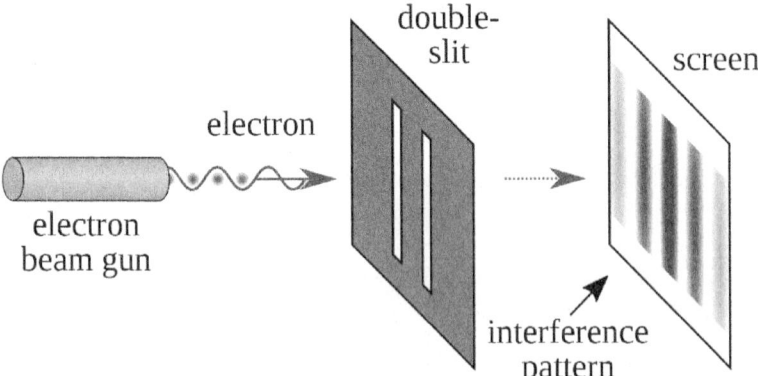

Albert Einstein was the first to propose that matter might behave both as waves and as particles in a serious way. As you learned in the last chapter, Planck unintentionally proposed it in his own work (Planck's constant) but didn't necessarily believe it himself. It took the work of several more scientists for the duality of wave-light particles to be fully accepted. Shortly after Max Planck made his discovery on the blackbody problem, a scientist named Philipp Lenard discovered that, when shining light at a metal, the frequency (color) of the light would change the *energy* of electrons ejected by the metal, and the amplitude (or power) of the light would change the *number* of the electrons ejected by the metal. However, he didn't necessarily know that he was working with photons; at the time, Lenard called them "light quanta" instead. At the time of his research, it still wasn't really accepted that light could function both as a particle and a wave, but his experiment irrefutably showed that light was functioning as a particle, not a wave, as it was believed to be at the time. Albert Einstein used this discovery, along with the work of Max Planck, to formulate his theory in 1905: the photoelectric effect. He related Lenard's discovery with Planck's constant, discovered several years earlier, and theorized that it worked the same way as the black-body problem: that the energy required to eject an electron from metal was quantized, and this quantification could be found using Planck's constant. This theory uses the same function we defined in the last chapter:

$E = hf$

Using this function, Einstein theorized that it was the frequency of the light (or the color) that was quantized, not the amplitude (brightness). According to classical physics, it was assumed that the brightness of the light would lead to more power, and would thus result in the release of more, or stronger, electrons. However, quantum mechanics found that that was not entirely the case. Simply adjusting the power of the light would not fix the problem – in the study, it was shown that while a low-powered, dim blue light (blue light has a high frequency) was able to energize and release electrons from metal, a high-powered, very bright red light was unable to do so (red light has a low frequency). This violated the laws of classical physics and led to the discovery that the frequency of light must be above a certain threshold (or quantum) in order to energize an electron.

Albert Einstein was the first to define "light quanta" as photons and the first to propose that they exhibited dual characteristics of waves and particles, but this wasn't until after the proposal of the photoelectric effect. A photon is one quantum of light. Einstein was also the first to propose that *light itself* is a quantized form of energy, rather than the release of energy from the object being quantized. However, Einstein's proposal was only a hypothesis at the time –

despite the support of experiments like those of Planck and Lenard, Einstein's proposal of the photoelectric effect was rejected for a number of years by the scientific community. They still believed, without more proof, that light was inherently a wave, not a particle like Einstein was proposing. It wasn't until another scientist by the name of Robert Andrews Millikan unintentionally confirmed that Einstein's theory was correct in 1915 that it started to be taken seriously. Einstein eventually received a Nobel Prize in 1921 for the discovery of the photoelectric effect.

We've been referring a lot to quantization in this chapter, but that's because the concept of quantization is intrinsically tied to the wave-particle duality of light and matter. Einstein's proposal of the photoelectric effect showed that light must function as a particle, but about one hundred years earlier, a scientist by the name of Thomas Young executed an experiment that irrefutably showed light to be a wave. This experiment was called the double slit experiment, and it's easy enough to replicate at home if you have the supplies.

In this experiment, Young placed a light source in front of a wall with two vertical slits cut into it. Behind that wall, he placed a screen, through which the viewer could see where the light from behind the wall would hit on the screen. The end result was a pattern of light that irrefutably showed light to be a wave – the pattern on the screen looked like a pattern of bars. This pattern is arrived at due to the nature of how light can function as a wave. Envision how, when you drop a pebble into a lake, ripple formations move out from where the pebble disturbed the water. Now, imagine dropping another pebble into the water alongside the first. Where the ripples between the two meet, some interesting things happen: where the crests of the ripples meet each other, taller ripples will be formed, and where the valleys of the ripples meet, deeper ripples will be formed. This is called constructive interference – the waves are meeting each other, but also boosting their strength at the same time. On the other hand is destructive interference, where one crest and one valley of the ripples meet, and this results in a cancellation of the wave altogether. On the screen at the end of the double split experiment, the constructive and destructive interference show up visually as alternating vertical bars, with black spaces in between them.

One of the most amazing results of the double-slit experiment – and one that undeniably proves that photons exhibit characteristics of particles and of waves – arises when you fire just one photon at a time into the double slit apparatus. When you do this, it would make sense to think that the photon would travel straight, and it would either bounce off the wall or go right through, correct? Well, this isn't what happens at all. Surprisingly, and almost unbelievably, when a photon is fired through the double slit apparatus, it will end up landing in one of the constructive interference areas of the screen on the other side. If you fire enough photons, you eventually will reform the waveform from the double slit experiment, too – the result is a series of vertical bars, just like before. The same result can be achieved not only with photons, but electrons, whole atoms, and even entire molecules.

Chapter 4: Quantum Non-Locality and the Bohr-Einstein Debates

Around this time, as you might imagine, things were changing very quickly within the scientific community. With the revelations that the world of quantum mechanics didn't really work at all like the great thinkers of classical physics had thought. Not only were there many uprisings, but also very many new thoughts were happening. It was a true age of discovery, and one could argue that it was the golden age of quantum physics.

Quantum Physics Made Easy

However, this influx of new thoughts and discovery also led to plenty of disagreements between scientists. For one, Niels Bohr, one of the great minds of the time, was one of the most vocal opposers to Einstein's proposal of the photoelectric effect – he staunchly refused to accept that it was true for twenty years after Einstein proposed it! This was good because more disagreement meant more experiments and more thought, but for a while, there was a lot of wondering whose ideas were best and whose were not. Bohr and Einstein, in particular, had such heated arguments over some concepts that their debates were immortalized in an article written by Bohr himself.

Einstein found it hard to accept the fact that quantum mechanics was almost entirely probability based, without any determinable explanation as to why. Einstein wanted to find the reason why quantum physics behaved the way that it did, and refused to accept that it was "just the way it was. "Specifically, he wanted to debunk the uncertainty principle and complementarity. When Bohr proposed a certain principle – the Copenhagen interpretation, a concept that partly endures today as one of the foundational concepts of quantum physics – Einstein was incensed. The Copenhagen interpretation said that, with its wave-particle duality nature, electrons were waves only until observed by an outside party, at which point they became particles. It was a complete violation of Einstein's belief in objective reality – he believed that the universe always existed in a stable state, independent of whether we observed it or not. The act of measuring an electron could also change its entangled electron, introducing the concept of quantum non-locality, which was anathema to Einstein's staunch belief in the locality. Locality is the principle that objects can only affect other objects that are in their local area, and that they cannot move faster than the speed of light. We'll be covering entangled particles more at the end of this chapter, and in chapter six.

Think of how a child is fascinated with a game of peek-a-boo. This is because young children haven't yet learned the human concept of object permanence – young toddlers and babies truly believe that, when their parent covers their face with their hands, their face is moving in and out of existence! This was exactly what Bohr was saying – the Copenhagen interpretation proposed that that was *actually* the way the quantum universe worked. Obviously, this had significant repercussions for the scientific community, and Einstein was intent on proving Bohr wrong on his interpretation.

The debates began when Bohr and a colleague, Werner Heisenberg (we mentioned him and his uncertainty principle in Chapter 2), announced that they both considered everything about the way quantum physics worked to be discovered. However, Einstein challenged this by presenting Bohr with several thought scenarios designed to challenge his ideas, leaving Bohr to solve them or risk his announcement being debunked.

Einstein's first challenge for Bohr was a modified version of the double-slit experiment from the previous chapter. It went as follows:

if another wall with only a single slit were to be placed in front of the double-slit wall, but the first wall was *on wheels* that were sensitive to very minute movements, then a photon bouncing off the first wall would move it slightly, allowing the experimenter to deduce which slit the photon would go through on the second wall. This result would violate the uncertainty principle.

Bohr refuted this by reminding Einstein that he hadn't debunked the uncertainty principle at all, but had instead neglected to consider the impossibility of measuring the movement of the first wall. In order to measure the amount that a photon would move it, you would have to be precise on a quantum level. At a quantum level, you must also follow quantum rules, and that means that Einstein would need to know the location and velocity of the wall before it was impacted by the photon. That would result in the violation of the uncertainty principle itself since one cannot measure velocity and location of an object at the same time on a quantum level.

Einstein had more tests for Bohr, however. Next, he gave Bohr a new experiment: if he had a box on a scale that had a photon inside of it, containing a built-in clock that would open the door for *just* long enough to allow the photon to escape, he would be able to measure the weight of the photon. If this was possible, that would mean that using Einstein's theory $E=mc^2$, Bohr could also find the energy of the photon. This also violates the uncertainty principle, since it also states that you cannot find the energy and time of a particle simultaneously. This seemed to throw Bohr for a loop, but he was again able to deny Einstein's scenario.

Bohr proved Einstein's scenario wrong by appealing to Einstein's own theory of relativity. First, Bohr cited Newton's first law. This law states that, for every action, there is an equal and opposite reaction. This means that, as the photon leaves Einstein's box, the box itself must also move away from the photon by an equal amount; however, it's impossible to find that amount because of the uncertainty principle. Next, Bohr stated that, according to Einstein's theory of relativity, the force of gravity will distort time as an object moves through space. Thus, it's impossible to tell if the time on the clock is correct, and as a direct result of that, one cannot tell the energy of the photon at a given time. Bohr: 2, Einstein: 0.

Long story made short, though, Einstein sought several times to disprove Bohr's theories, but every time, Bohr was able to shoot Einstein's rebuttals down. Einstein and Bohr continued to debate for the rest of their lives (they were friends), and Bohr's ideas are still considered to be true for the most part by today's scientists.

However, there was one last concept proposed by Bohr that Einstein simply could not accept – one that bothered him more than any other. So much so, that Einstein himself mockingly called it "spooky action at a distance," rather than the name Bohr gave it: quantum entanglement.

Chapter 5: Quantum Entanglement and Teleportation

Quantum entanglement is indeed spooky – that's something we can't argue with. Truly, it often seems more like magic than science, but then, many parts of quantum physics wind up feeling this way. We still have much to learn and explain about the world, and quantum entanglement is among the most mysterious and least understood.

When two particles are entangled, it means that any change to one particle in the pair, no matter where that particle is in the universe, will result in compensation by the other particle. According to Einstein's theory of relativity, the light was the fastest thing in observable existence, so this bothered him immensely. If, for example, a change in one entangled atom was made on Mars, and the other atom on Earth instantly shifted to compensate, this would be many times faster than the speed of light.

To understand quantum entanglement, you must first be introduced to the concept of spin. For ease of understanding, let's assume that all particles in the universe are spinning up or down and that the universe must always have a net spin of zero (that means there must be an equal number of particles spinning upwards and an equal number spinning downwards). If by chance, someone changes the spin of one of these particles to be the other direction, the other entangled particle, wherever it is in the universe, will switch directions to compensate. This becomes even weirder when noted that, when we measure the spin of a particle, it will change *simply because we measured it* (this harkens back to the Copenhagen interpretation). However, the spin of one entangled particle will *always* be opposite of its partner particle. This means that we can essentially change the nature of a particle across the universe instantaneously, simply by measuring the spin of its entangled particle.

John Bell proved this irrevocably in 1964 when he introduced one of the greatest theories of quantum physics: Bell's Theorem. Einstein wanted to believe that these entangled particles, rather than relying on "spooky stuff," had hidden information incorporated into them during their creation that would determine the way they would spin when measured. However, Bell's theorem irrevocably proved this to be false using simple math. Without getting too deep into the math, Bell essentially tested out all the scenarios that could arise from these particles having "hidden information" within them and solved for the math that would result from these measurements. What he came up with was that, no matter what was done, no matter how many times he measured it, there was no way that the math would allow for hidden, local information within the particles; rather, quantum entanglement was the only acceptable solution to Bell's Theorem.

Interestingly enough for us, teleportation is theoretically possible through the use of quantum entanglement. Physicists have used, to date, quantum entanglement to teleport atoms, electrons, and photons instantaneously between one place and another – so, someday, teleportation may be possible for us, as well. Essentially, in order to teleport a particle via this process, it needs to be combined and entangled with an entangled particle in a specific way. Using a lot of complicated math, and essentially a process that would involve "measuring" the other particle in order to instantaneously teleport the extra particle, it is possible to do, and it has been done. However, it's extremely difficult, currently, with particles much bigger than an atom, and only works across limited distances. For the moment, it's still not possible for us to teleport people – or living things, for that matter. It may be possible someday, though.

Chapter 6: Quantum Superpositions and Schrödinger's Cat

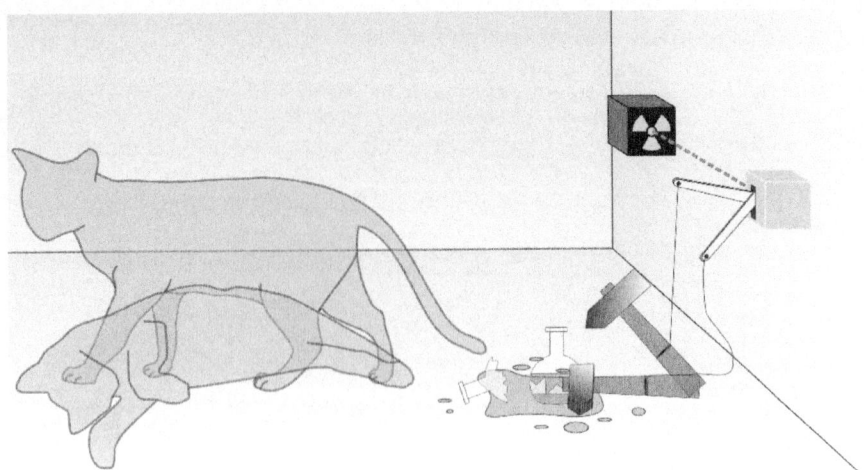

Quantum superposition is another foundational aspect of quantum physics that has no bearing – and even sounds ridiculous – through the lens of classical physics. Quantum superposition is the ability of a quantum object to be in two states at the same time: that of a wave and a particle. As we ascertained before, it's not really possible to see the wave nature of large objects (such as a car), because the wavelength (or frequency) of an object goes down as the velocity (or momentum) goes up. Since a car has a relatively large weight and size, its wavelength is very small. Take a photon on the other hand, which is very small and has virtually no mass: its frequency is quite visible. Remember the dual slit experiment that we mentioned earlier? Well, if we're looking at a photon being fired singularly into the dual slit apparatus, it's not actually picking one slit to go through and end up

Quantum Physics Made Easy

on the other side. On the contrary, according to quantum physics, the photon is superimposed into both sides at the same time due to its wave-particle nature. It goes through *both* slits, and this is what results in the waveform pattern at the end of the experiment. We know this because, if you cover up one of the slits, the vertical bar pattern goes away.

Schrödinger's cat is one of the most popularized scenarios to demonstrate quantum superposition, but it's lost some of its original meaning due to its liberal use. In it, the scientist Erwin Schrödinger proposes that if a cat is sealed into a box with a device that has a fifty-fifty chance of killing the cat in one hour, that, at the end of that hour, the cat is both alive and dead until the box is opened. However, many people who cite the paradox don't know that the device is actually rather complicated in the original analogy – it's not just a bomb that kills the cat, for example. The device is actually a Geiger counter (a device that measures radiation), a vial of poisonous gas, and some radioactive material that has a fifty percent chance of decaying within the hour. If the radioactive material does decay within the hour, the Geiger counter will trigger a hammer that will shatter the vial of poisonous gas, killing the cat. If the radioactive material doesn't decay, the cat will be just fine when the box is opened.

In the scenario, according to quantum physics, until it is observed, the radioactive material is both in a decayed and non-decayed state at the same time, since it is superimposed to both states. This would also mean that the cat must be both dead and alive at the same time, but Schrödinger didn't consider this to be possible. Schrödinger's point was not that the cat was both dead and alive, but that it's very difficult to reconcile quantum physics with classical physics on a large scale in our day to day lives. A cat can't go through two slits at the same time as a photon can, and since superposition tends to break down when we try to measure or observe it, that means we'll never know the status of the radioactive material and the cat until the box is opened. The entire point of Schrödinger's experiment was to show that the Copenhagen interpretation was ridiculous. Technically, if you think about it, the cat could observe the radioactive material while it's inside the box, rendering the superposition useless anyway…right? For reasons like this, the Schrödinger's cat scenario is often described as a paradox – however, physicists of today agree that despite Schrödinger's contemptuous attitude towards the paradox when he created it, the simultaneous "dead and alive" state of the cat is actually quite possible. Scientists today largely agree that quantum superposition does exist among objects on a quantum and non-quantum scale. One of the alternative explanations to the Schrödinger's cat problem that supports this is the infinite world's theory: this theory states that when the box is opened at the end of Schrödinger's experiment, two worlds are created: one where the cat is alive, and one where the cat is dead. According to this theory, we're simply part of the timeline where one result happened, and there is a copy of our timeline living out the opposite result. However, this would theoretically lead to infinite timelines and universes, so the many worlds theory is not widely accepted.

One of the more accepted solutions to Schrödinger's paradox is called "decoherence." This theory essentially says that the universe chooses which outcome will happen based entirely on chance when there is an interaction between quantum and non-quantum-sized objects – like a quantum roulette, so to speak – and then merges the result into the current timeline. This was an interpretation that Einstein took extreme offense to, as he was a staunch believer in cause and effect. However, there's no more evidence for decoherence than there is for the many world's theories. According to the evidence we have, both are equally likely.

Schrödinger's cat, while easily his most well-known contribution to quantum physics, was not his only one – and, arguably, it wasn't his most important, either. In 1925, Schrödinger introduced the Schrödinger equation (creative name, right?). His equation looked like this:

$$H(t)|\psi(t)\rangle = i\hbar\frac{\partial}{\partial t}|\psi(t)\rangle$$

Don't worry too much about this one, as it's a rather complex equation, and we won't be analyzing it here. Essentially, the equation tells us the probability of finding a particle at any specific location within its waveform. Theoretically, it tells you the most likely place where the particle will be according to the variables in the equation.

Chapter 7: String Theory and the Theory of Everything

There's a good chance that, as a reader of this book, you've heard of the term "string theory," but you have no idea what it means. As it turns out, this is the case for many people, and not by their own fault, either – string theory is just really, really confusing.

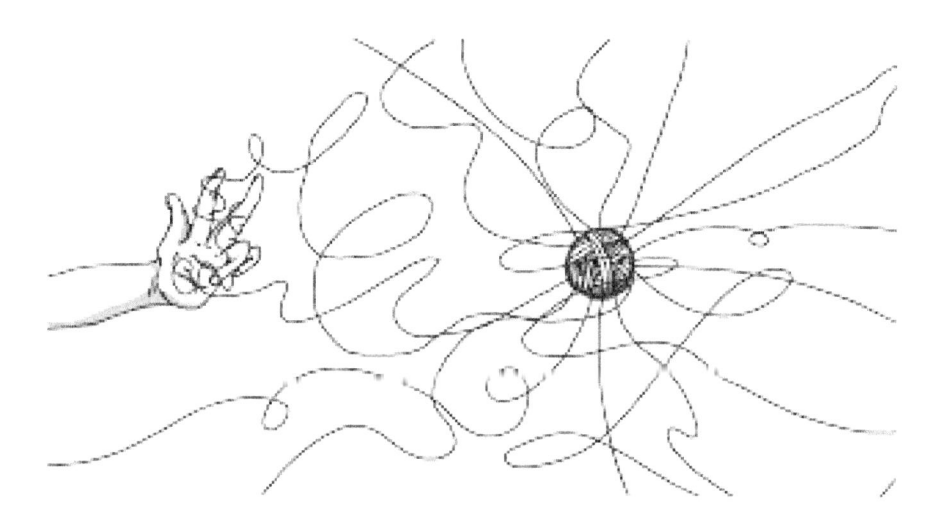

String theory is so named because it was created to describe the nature of quantum objects as something like a vibrating string, and those objects were measured based on those vibrations. Without getting too deeply into it – string theory involves a lot of work with subatomic particles, which are the smaller particles that makeup atoms (like quarks, for example) – string theory was essentially formulated to help explain strange interactions that sometimes happened between some of the known subatomic particles. These particles would sometimes act like they were bound together by strings. Thus, string theory was formed.

Now, one of the most interesting parts of this early string theory was that these weird, string-based interactions worked out some really mysterious math. For one, the vibrations in the strings predicted the existence of a certain particle: the graviton. It's the only quantum theory that has been able to successfully do this so far. Theoretically, a graviton is a particle that causes gravity, and one of string theory's biggest advantages here is that its graviton is actually shaped like a *donut*. That may sound strange, but that same shape actually prevents many of the mathematical anomalies that arise when trying to envision gravity as a particle. However, with today's science, we still can't prove that the graviton even exists – gravity still presents many mysteries to us.

String theory was able to successfully describe gravity as its own quantum particle this way. However, string theory has many limitations that keep it from being an attractive (or logical) solution to the mysteries of the quantum universe. For one, the final proposal of string theory requires *ten dimensions* to function properly, but we've so far only observed four dimensions in our existence. To remedy that, physicists and scientists alike have tried to make string theory work in its prerequisite ten dimensions, then remove the extra dimensions that don't apply to our universe. However, none have yet been successful. In addition, there are several other versions of string theory – there isn't just one! Bosonic string theory was one example, and it required *twenty-six dimensions*.

In 1995, a man by the name of Ed Witten managed to combine many of the varied string theories together into one, which he dubbed M-theory. However, M-theory required eleven dimensions, so while it managed to unify many of the past string theories into one neat package, it still didn't seem anywhere near reasonable science. (It's also very complicated, and unnecessary for us to analyze here.)

This doesn't mean that string theory isn't useful. There's also a concept out there called the "theory of everything," and, as the name implies, this would be an all-encompassing theory to explain how quantum physics and quantum particles worked. Although it's one of the premier goals of today's physicists and scientists, we haven't formulated a working theory of everything yet. However, string theory makes a compelling case for becoming or at least supplementing, a potential theory of everything.

For one, string theory makes a very neat argument for explaining the nature of particles. Envision a guitar string in slow motion: the string moves in a wave shape, just like light, or sound. The tension (how taut the string is pulled) of the string also controls what note any string on a guitar plays. If you pluck the string, it will vibrate. This is the prevailing theory behind the strings: each string has a different frequency (like the note of a guitar string) that it vibrates at, and each string also has a different length, which changes the number of notes that the string could possibly play. Think about how, when playing the guitar, moving your fingers down the neck, shortening the string, will play higher notes while moving your hand up the neck will lengthen the string and play lower notes. Interestingly enough, the intervals between these possible "notes" are also defined by a familiar variable: Planck's constant.

Unfortunately, as convenient as this might seem, there's one question that scientists seem to be unwilling (or unable) to answer about string theory. What are the strings made of? Answers that have been tried are "pure mass" (whatever that means), irregularities in the fabric of reality, and the "energy" of "existence." The best answer, however, is, "The answer is irrelevant because they're *there*."

One of the biggest downfalls of string theory is that, because it's so complicated and exists in so many additional dimensions (that we can't even find), we can't test it. However, we also can't rule it out as entirely wrong, either. This creates a bit of a catch twenty-two: should scientists and physicists keep pouring man hours and research into a theory that we technically can't even test, or should we just give up on it, and risk sidelining a potential theory of the universe? There's no good answer to this question, but it's one of the pitfalls that plague string theory and those interested in it.

Despite its shortcomings, though, many scientists still believe that string theory is the way things are – or, at least, it's a step to figuring it out. However, until we find six or seven more dimensions, we'll have to wait on that.

Chapter 8: Black Holes and the Mystery of Quantum Gravity

We talked about Einstein's theory of general relativity in the introduction chapter. However, Einstein's theory is technically unreconcilable with quantum physics, for several reasons that we've mentioned in this book. According to Einstein, black holes are regions of ultra-intense gravity within spacetime that are so powerful, they pull even light inside.

The idea of a black hole was first proposed by a clergyman named John Mitchell in 1784. His ideas were dismissed for the most part because, shortly afterward, the light was "discovered" to be a wave instead of the particle (at the time). Scientists were therefore unsure if gravity would be able to act on light "waves" instead of "particles," and thus largely forgot about the concept. That is until Einstein brought them back in 1915 with his general theory of relativity. From there, exactly how black holes are formed, how they work, and how they influence space and time have been hotly debated by many physicists. Scientists do believe the time is so distorted in a black hole that, to an outside observer, time would appear to stop inside of it. However, if a person were to fall into a black hole, time would proceed for them normally (resulting in a rather gruesome death).

As they destroy things, black holes are believed to completely delete them from the universe. However, this idea is a violation of quantum laws, which say that the information in quantum particles can never be fully destroyed. This is solved in part by a proposal by Steven Hawking – another name you should know – who predicted that black holes give off radiation as they shrink, in which this information is stored and thus released back into the universe.

Black holes can be created in a few different ways. One way would be by violating Planck's law. This has to do with the Planck constant, which you learned about before. Planck's constant also defines several measurements within quantum physics, and these measurements denote certain limits of measurement that can be achieved within the quantum universe. For example, if you tried to measure a particle's position with a laser at a greater accuracy than one Planck length (1.6×10^{-35} meters, which is very, *very* small), the power of the laser would end up creating a very small black hole. Ironically, the black hole created would be exactly the size of one Planck length. Since time and space are intertwined, a black hole can also be created when trying to measure a length of time less than one unit of Planck time (10^{-43} seconds, which is *very short*).

Black holes are also, more traditionally, created by the collapse of very large stars. At the end of a star's life cycle, when the fuel inside the star has all but run out, the mass of the star itself causes it to collapse, and all the matter inside of the star get sucked in. Sometimes, this can create smaller, dimmer stars, or it can create quasars (a type of ultra-bright, fast spinning, a very small star that we touch on in the extra facts chapter, also containing a black hole). Remember Heisenberg's uncertainty principle? Theoretically, this also can apply to the creation of black holes. If you'll remember, according to the uncertainty principle, it's impossible to determine a quantum particle's momentum and position with high precision at the same time, and if you attempt to measure one or the other more precisely, the other measurement becomes less and less precise. If you get precise enough – say, down to one Planck length or less – the other value becomes so large that the particle is mathematically capable of turning into a black hole. This doesn't mean the particle is *actually* becoming a black hole, but more of a math problem illustrating the difficulty of measuring quantum particles. This is called an absurdity or a nonsense result – according to physicists, this means that something is missing from the math. They're missing a formula, a variable, or something else. This is the mystery of quantum gravity – even today, we're still missing something in the math required to make gravity (and, by extension, black holes) make sense.

These issues are propagated by the fact that there is no way to measure gravity on a coordinate plane. Since gravity exists within four dimensions, it can't be figured out using the two-dimensional math that scientists have traditionally used to figure out other quantum theories. Additionally, if you try to apply four-dimensional gravity to these two-dimensional theories, you receive anomalous answers – the math just doesn't make sense. Essentially, gravity *is* space-time. As a result, you can't anchor any math within space-time to help figure it out.

At its core, the reason why this math doesn't work is because, as we mentioned at the beginning of the chapter, Einstein's general theory of relativity is incompatible with quantum mechanics on a basic level. The general theory of relativity works at a level that we can see, but at the quantum level (i.e. below the Planck scale), it unravels. This brings us full circle because that's why quantum mechanics was created in the first place; however, quantum mechanics has yet to account for the anomaly of gravity. For now, physicists continue to assume that we simply haven't found the right theory to solve it yet.

(As we mentioned in a previous chapter, many scientists believe string theory may be one of the candidates to do this.)

As we mentioned a little earlier in this chapter, according to Einstein's theory of general relativity, black holes are essentially anomalies in space-time that are holes with infinite mass. Black holes (and everything in the universe that experiences gravity, to a lesser extent) experiences something called gravitational time dilation. This means that, the closer you are to the center of the black hole, the slower time moves to an observer outside of the black hole. However, strangely enough, time will proceed at a normal pace to those closer to the black hole. This means that, if a pair of twins were to stand by a black hole – one twin very close to the black hole, and the other very far away – when the twins both came back to earth, the twin who was further from the black hole will have aged more than the one who was close to the black hole. Crazy, right? This also happens, to a lesser extent, to the satellites orbiting the earth, as well as the international space station (the ISS). Clocks that orbit Earth from space will slowly move ahead of clocks on earth.

Even stranger – continuing with the twin metaphor – if one of the twins were to actually fall *inside* of the black hole, and the other twin was to watch, time would appear to *completely stop* on the twin inside the black hole as the twin crossed over the edge: the twin watching outside the black hole would just see their other twin stop, freeze totally in time, then never move again, no matter how long the other twin waited for them. No one really knows whether it would be possible to escape a black hole after this had happened (according to all known laws of physics, any *normal* person going into a black hole would be crushed by the incredibly strong gravitational force long before they made it to the center, of course), if it were possible to survive the experience. One theory is the white hole theory, which says that for every black hole, there is a matching white hole somewhere else in the galaxy that spews out all of the matter consumed by the black hole. If the person who fell into the black hole survived, then according to this theory, the black hole would work as a teleportation device. However, although we have documented cases of proposed white holes in the universe (the white hole GRB 060614 was found in 2006), there's no way to prove that these holes are in any way connected to black holes.

Some scientists have proposed that the Big Bang that created the universe was actually a white hole. The same paper that proposes this also says that white holes should be spontaneous, limited occurrences, rather than long-lasting singularities like black holes.

Chapter 9: Other Interesting Facts

Pilot Wave Theory

According to accepted ideas of quantum physics, the wave portion of wave-particle duality is essentially a cloud of probability of where the particle might be – only when the particle is measured does it take on a specific location. However, there is an alternative and very compelling theory that actually describes the wave-particle duality of being truly physical and observable. It's called the pilot wave theory, and it was proposed by Louis de Broglie in 1927. According to his theory, a physical wave is always guiding a physical particle that always has a definite location – no probability involved. He states that, if it were possible to measure both the position and velocity of a particle at the same time, it would be possible to figure out its entire trajectory, based on the way the wave would guide it. The progression of these waves was also predicted by Schrödinger's equation, to boot. However, de Broglie essentially abandoned the theory in favor of the Copenhagen interpretation until David Bohm rediscovered the theory in 1952. Despite being as feasible as the Copenhagen interpretation, it's not nearly as popular or widely embraced.

Quasars

Quantum Physics Made Easy

Quasars are nearly as mysterious (and nearly as cool) as black holes. The first quasar was discovered in 1962 when the moon passed in front of an as yet undiscovered quasar, helping the scientists of the time pinpoint it in the sky (and realize that it wasn't just any normal star). Scientists, realizing that something was different, did a spectrum analysis of the light coming from the star (in layman's terms, they analyzed what wavelengths the star was giving off). Strangely, the light coming from the quasar was nothing like the light from a normal star. One of the most striking observations made was that the light from the quasar was significantly redshifted (we touched on this very early in the book). Redshift means that the photons coming from the star began to stretch due to gravity from their long travel through the universe, and the light of the star turns redder than it normally would be. This was strange to scientists because the quasar was far too bright to be as far away as the redshift seemed to indicate; any normal star would be much dimmer, or nearly invisible. That meant that the star they were viewing was impossibly bright – much more so than any other star they'd ever seen. It took around twenty more years for scientists to figure out what exactly the super-bright star actually was. Quasars are actually supermassive black holes, which we believe have been created by galaxies colliding with one another and merging. Some of the matter within these colliding galaxies is sucked into the black hole, moving faster and faster around it (visualize how water turns into a vortex as it's sucked down a drain) until it eventually burns up, radiating light and heat. That's why these quasars are so bright. Scientists also now believe that every large galaxy has a supermassive black hole at the center of it, with the potential of creating one of these quasars. Many quasars also, for reasons that scientists haven't fully figured out yet, tend to shoot streams of gas and matter up and down, away from the black hole, at incredible speeds. The whole phenomenon ends up looking something like a spinning top, with streams at the top and bottom, coupled with a wide, flat disc around the center. Scientists also believe that quasars were one of the essential phenomena that helped create our galaxy and many others.

Mini-Facts

- Quantum superposition is also what allows certain compounds to conduct or block electricity, and it is what enabled computers and electronics to function as they do today. If quantum superposition didn't exist, we wouldn't have microchips, cable, or wired electricity!

- In 2015, scientists were able to photograph (and therefore observe) the function of light as both a particle and a wave for the first time. This picture is available for viewing on the internet.

Quantum Physics Made Easy

- On February 11th, 2016, scientists successfully detected gravitational waves for the first time, believed to be from the merging of black holes.

- One of the reasons why scientists believe we haven't found any alien civilizations is because of the nature of how light and sound waves travel in space. Even though light travels extremely quickly, it takes it a significant amount of time to reach far regions of the universe (hence the term light-year, which, predictably, is a measurement of how fast light will travel in a year in space. A light-year is almost six trillion miles). Because of this, when we peer into space through a telescope, we can only see the light that is reaching us at that given time. So, for instance, if we're looking through a telescope that is pointed at a planet ten-thousand light-years away, we're actually seeing what the planet looked like ten-thousand years ago since the light from that time would be just reaching our telescope. With our current technology, we have no way of seeing what is actually going on *right this minute* when it's that far across the universe.

Conclusion

Thank you for making it through to the end of *Quantum Physics Made Easy* – lets hope it was informative and able to provide you with all of the tools you need to achieve your goals, whatever they may be.

Quantum physics is an endlessly fascinating subject and one that we think everyone should learn about in some capacity. After all, as we discover more and more about quantum physics, more of the future open to us – things like teleportation, supercomputing, and ultra-fast space travel all feel like they're just around the corner if we can finally unravel how quantum mechanics works. We're not all scientists, of course, and we can't all conduct experiments in order to figure out the next big discovery, but we can all do our part just by learning a little bit about how the universe around us works.

There are so many things going on in the universe around us that we can't explain. However, there's beauty in this un-knowing, and there's an indelible fascination in the discovery of the world around us. How boring would it be to already know exactly how everything around us worked? If we knew that, we would be able to predict everything, from whether there was alien life in the far reaches of the universe, to when our own planet and solar system would die. While some might argue that it might be better to know, the vastness of the universe allows for infinite new things to discover, crazy new laws and theorems to figure out, and new phenomena just over the event horizon.

The next step is to incorporate the knowledge you've gained here into your everyday life. Whether you use your new knowledge simply to brag, to help others, to do your own research, to deepen your understanding of our world, or to become the next Albert Einstein, we're sure it will enrich your life in new and exciting ways. Please never stop reading, and more importantly, never stop learning.

Finally, if you found this book useful in any way, a review on Amazon is always appreciated!

Donald B. Grey

Quantum Physics Made Easy

Connect with us on our Facebook page www.facebook.com/bluesourceandfriends and stay tuned to our latest book promotions and free giveaways.

Printed in Great Britain
by Amazon